SADDLEBACK
EDUCATIONAL PUBLISHING

think green

# Global Warming

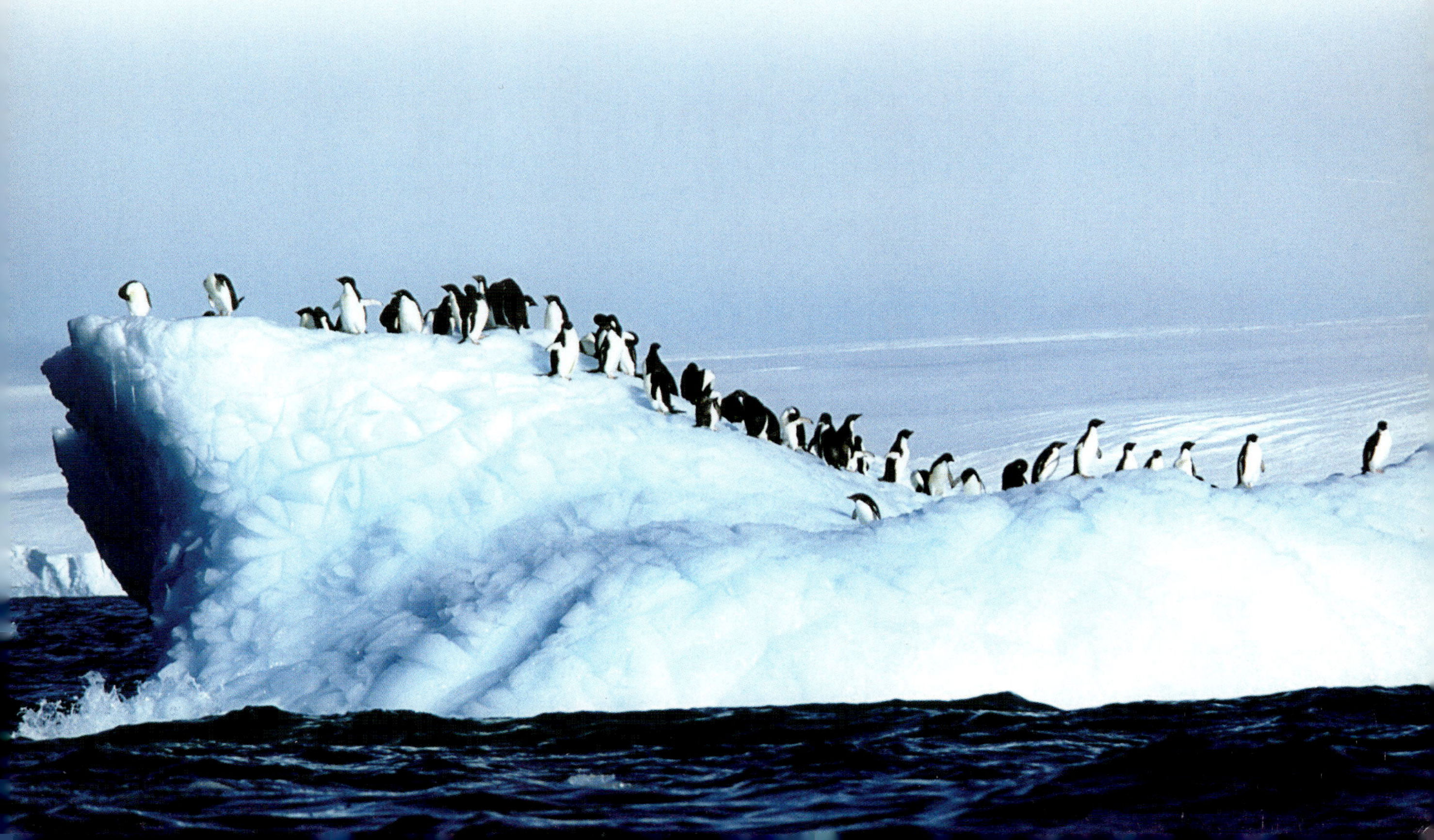

Three Watson
Irvine, CA 92618-2767
Website: www.sdlback.com

**Copyright © 2010 by Saddleback Educational Publishing**
All rights reserved. No part of this publication may be reproduced, stored in a retrieval system, or transmitted in any form or by any means, electronic, mechanical, photocopying, recording, or otherwise, without the prior permission of the publisher.

ISBN-13: 978-1-59905-455-1
ISBN-10: 1-59905-455-8
eBook: 978-1-60291-781-1

Printed in China
15 14 13 12 11   1 2 3 4 5

# Contents

Global Warming: An Introduction ........ 4

Earth Is Getting Hotter ........................ 6

The Greenhouse Effect ...................... 8

Greenhouse Gases ........................... 10

Ozone Layer Depletion ..................... 12

The Carbon Cycle ............................. 14

The Kyoto Protocol .......................... 16

Climate Change ................................ 18

Severe Weather ................................ 20

Heat Waves ...................................... 22

Earlier Spring ................................... 24

Heavy Rain ....................................... 26

Heat Harms the Atmosphere ............. 28

Heat Increases Hurricane Strength .... 30

Arctic Ice Melting ............................. 32

Worldwide Pollution ........................ 34

Air Pollution .................................... 36

Driving ............................................ 38

Public Transportation ...................... 40

Permafrost ...................................... 42

Green Homes ................................... 44

Collaborative Housing ..................... 46

Deforestation and Altering
    the Carbon Balance ..................... 48

Fossil Fuels ..................................... 50

Ocean Acidification ......................... 52

Population Shifts of Plants
    and Animals ................................ 54

Fire .................................................. 56

CFLs ................................................. 58

Indoor Air Pollution ......................... 60

Facts and Figures ............................ 62

Index ............................................... 64

Glossary .......................................... 65

**THINK GREEN:** Global Warming

# Global Warming: An Introduction

Earth gets most of its heat from the sun. The sun gives off heat and light, which are energy. Some of the sun's energy reaches Earth. Some of this energy is *reflected*, or bounced back, into space. Earth's surface also absorbs and stores some of the energy as heat. When Earth's surface and atmosphere become warm, they give off, or *radiate*, heat back into space. This helps keep Earth from becoming too hot or too cold.

## What is global warming?

*Global warming* is the increase in the average temperature of Earth's surface. Earth has warmed by about 1°F over the past 100 years. Many scientists think that Earth is warming because of things humans have done. A warmer Earth means more heat waves, rising sea levels, floods, droughts, and wildfires. A warmer Earth could also mean more illnesses that spread to many millions of people. These disease events are called *epidemics*.

## Causes of Global Warming

### Human Actions

The *atmosphere* is made up of gases. Some of these gases trap the sun's energy and keep Earth warm. They are called *greenhouse gases*. If there are more greenhouse gases in the atmosphere, more heat is trapped. This makes Earth warmer. Burning coal, oil, and natural gas gives off a lot of greenhouse gases. Cutting down large areas of trees is called *deforestation*. Deforestation increases greenhouse gases. Other activities also increase greenhouse gases.

### Volcanic Eruptions

Volcanoes release a lot of water and carbon dioxide when they *erupt*. When volcanoes erupt, they explode lava and other material into the air. Over thousands or millions of years, many eruptions of volcanoes can raise carbon dioxide levels. This carbon dioxide is enough to raise Earth's temperature.

### Changes in the Sun

The sun is the main source of energy for Earth's climate system. Small changes in the sun's energy over a long time can lead to climate changes. Some scientists think that some warming is due to more energy coming from the sun.

Global Warming: An Introduction   5

## Effects of Global Warming

### Less Water
Changes in where rain falls, and how much rain falls, are an effect of global warming. Less rainfall in many places has led to less ground water. Rising sea levels affect water in coastal areas. The fresh water mixes with salty seawater and becomes polluted. This causes shortages of water used for drinking and to water crops.

### Failing Crops
Long dry seasons can lead to crop failures. Crop *yield* is the amount of food produced by an area of planted fields. Cereal crop yields are expected to drop in Africa, the Middle East, and India.

### Changing Weather Patterns
Global warming has led to climate changes. Changing weather patterns have affected plants, animals, and people in many parts of the world. According to the World Health Organization (WHO), 5 million people become ill and 150,000 people die every year because of climate changes.

### Less Sea Ice
Arctic sea ice continues to disappear because of global warming. At the current rate, by 2060 there will be no Arctic sea ice.

### Effect on Animals
The population of many kinds of polar animals has decreased. The polar bear population in Canada's western Hudson Bay has dropped by 22% since the late 1980s. Polar bears are also smaller than they used to be. The North Sea codfish is also smaller and produces fewer offspring.

> **Did you know?**
>
> Computer models have calculated that the Earth's average temperature could rise as much as 4°F to 11°F by the end of the 21st century.

**THINK GREEN:** Global Warming

# Earth Is Getting Hotter

Earth is getting hotter. Some scientists believe this is because the sun is burning more brightly. Other scientists have shown that larger amounts of greenhouse gases are making Earth hotter. This heat is dangerous to humans and animals.

# Earth Is Getting Hotter

## Diseases

Increasing temperatures have led to the outbreak of diseases such as malaria and dengue fever. These diseases are appearing in areas where they had not been before. Warmer temperatures mean that the animals and insects that carry disease can live in more places.

## Extreme Weather

*Precipitation* (rainfall and snowfall) has increased worldwide. Heavy storms have led to floods and landslides. Hard rain also causes soil to be lost by washing away, or *erosion*. Rising temperatures have also dried out soils much more than usual during the summer. Droughts have become widespread. The risk of wildfires is higher.

## Bleaching Reefs

Coral reefs are home to many kinds of animals and plants that live in no other places. Coral reefs are experiencing *bleaching* because the algae on which corals live is disappearing.

## Rising Sea Level

Warmer temperatures have caused glaciers and ice shelves to melt. Melting ice makes the ocean water rise on coasts. Global sea levels have risen by 4 to 10 inches over the past 100 years.

**THINK GREEN:** Global Warming

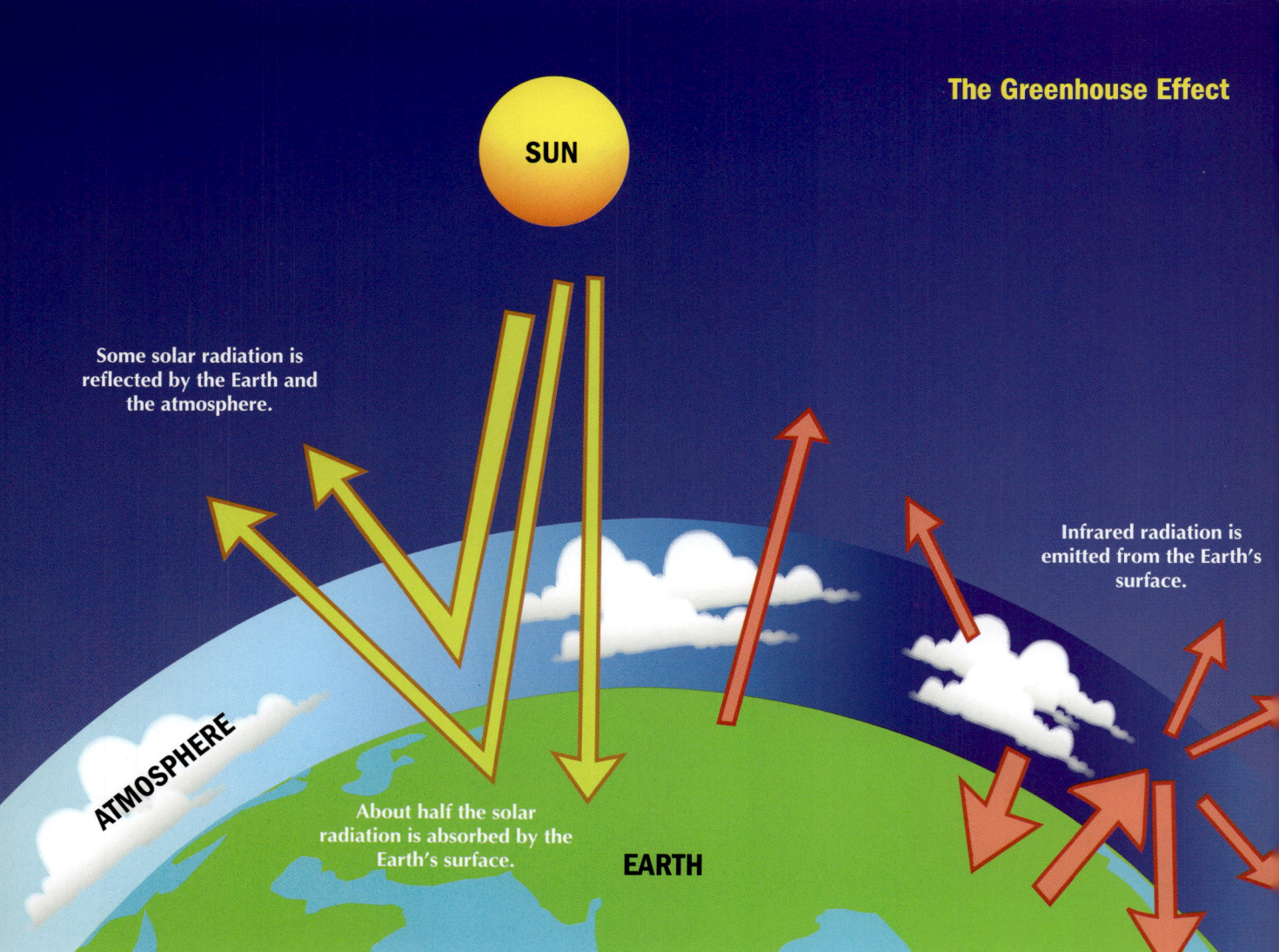

# The Greenhouse Effect

Earth is surrounded by a layer of gases known as the atmosphere. Greenhouse gases in the atmosphere act like glass in a greenhouse. They let the sun's energy pass through, but stop some of the heat from leaving Earth's atmosphere. The way they act to trap heat is called the *greenhouse effect*. The greenhouse effect is very important. Without it, the temperature of Earth would be too cold for most life.

# The Greenhouse Effect

### Did you know?
On Venus, the greenhouse effect is so strong that the surface temperature can reach 900°F.

## Effects of Enhanced Greenhouse Effect
Enhanced greenhouse effect is responsible for global warming. The effects of enhanced greenhouse effect are:
- Climate change
- Melting glaciers and rising sea levels
- More rainfall and floods in certain areas
- Severe droughts leading to hunger and death
- Many kinds of plants and animals dying out completely, or becoming *extinct*

## Enhanced Greenhouse Effect
The *enhanced greenhouse effect* is the increase in the natural greenhouse effect caused by human actions. The enhanced greenhouse effect traps extra heat. This increases Earth's surface temperature. Enhanced greenhouse effect is an outcome of many different human actions.

## The Glass Greenhouse
A greenhouse is a building made of glass. Plants are grown in greenhouses. The glass in a greenhouse lets in light energy from the sun. This energy warms the air inside the greenhouse. The warm air inside cannot escape through the glass. This keeps the greenhouse warm and protects the plants from the cold outside.

# Greenhouse Gases

Some greenhouse gases are natural. Others, like *chlorofluorocarbons*, are made by people. The six main greenhouse gases are water vapor, carbon dioxide ($CO_2$), methane ($CH_4$), nitrous oxide ($N_2O$), ozone ($O_3$), and chlorofluorocarbons.

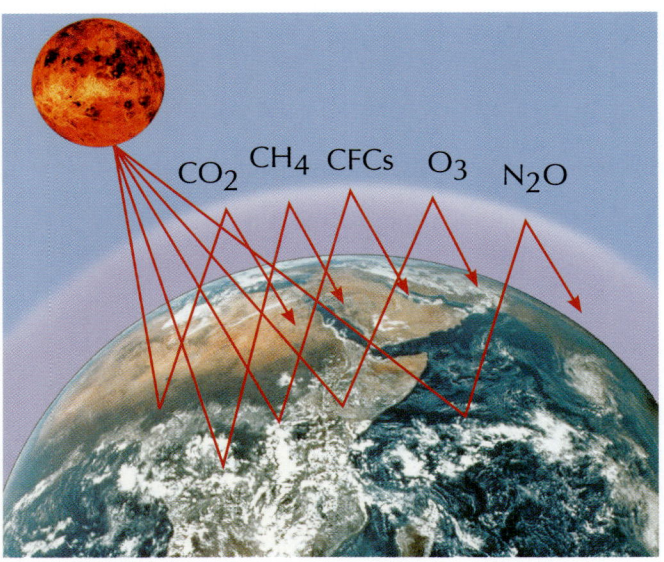

## Water Vapor

Water vapor is water in gas form. It is the most common greenhouse gas in the atmosphere. As global temperature rises, more water *evaporates*, or is heated and turns into gas. Warm air can hold more water vapor than cold air can. Higher amounts of water vapor absorb more heat. This causes global warming.

## Carbon Dioxide

Carbon dioxide is an important greenhouse gas. It is the main gas responsible for the enhanced greenhouse effect. Human actions have produced more carbon dioxide in recent years. Carbon dioxide is given off by the burning of fossil fuels. The amount of carbon dioxide in the atmosphere has increased since the Industrial Revolution in the 1700s.

## Discovering Greenhouse Gases

In the 1800s, scientists discovered how the greenhouse effect worked. They suggested that burning *fossil fuels* could cause Earth's temperature to change. Fossil fuels are oil, coal, and natural gas. Fossil fuels take a long time to form deep in the ground.

Greenhouse Gases 11

## Methane

Methane is a natural greenhouse gas. There is less methane in the atmosphere than carbon dioxide. However, methane is 21 times better at trapping heat than carbon dioxide. Methane is given off into the atmosphere both naturally and from human actions. Methane is the fastest increasing greenhouse gas.

## Nitrous Oxide

Nitrous oxide is one of the three main greenhouse gases. It can trap about 300 times more heat than carbon dioxide. Using more fertilizers in farming that contain nitrogen, burning fossil fuels, and making some chemicals can release nitrous oxide into the atmosphere.

## Tropospheric Ozone

Ozone is natural, but can also be created by humans. It is found in both the *stratosphere* and *troposphere*. These are two layers of Earth's atmosphere. Ozone in the stratosphere, very high above the Earth, is known as the *ozone layer*. It protects us from the harmful ultraviolet rays of the sun. But ozone in the troposphere, near the ground, acts as a greenhouse gas. This ozone in the troposphere is often called "bad" ozone. It adds to smog, which can cause health problems. Ozone is limited to cities and industrial areas. Ozone in the troposphere has increased by about 30% since the Industrial Revolution.

### Did you know?

Carbon dioxide can remain in the atmosphere for 50 to 200 years.

## Chlorofluorocarbons

Chlorofluorocarbons are greenhouse gases that are made by humans. They contain carbon, chlorine, and fluorine. They are used in aerosol cans (like the ones that contain hairspray) and air conditioners. Chlorofluorocarbons destroy stratospheric ozone. People are using fewer chlorofluorocarbons to protect the ozone layer.

## How can we decrease greenhouse gas emissions?

- Plant more trees.
- Turn off lights and other things that use energy when not in use.
- Use fluorescent light bulbs.
- Use alternative fuels to run your cars or trucks.
- Make less waste. Buy recyclable products.

**THINK GREEN:** Global Warming

# Ozone Layer Depletion

The sun provides us with light and heat. The sun's energy also contains *ultraviolet radiation*. This type of energy is harmful to humans and other life. The ozone layer absorbs most of the ultraviolet radiation from the sun. The ozone layer is disappearing in many places. Some greenhouse gases, such as chlorofluorocarbons, are destroying the ozone layer. As a result, harmful radiation from the sun reaches Earth.

## What does the ozone layer do?

The ozone layer stops the harmful radiation from the sun. Ultraviolet radiation can cause skin cancer, cataracts, and snow blindness. It can also affect some crops as well as life in the oceans.

## Chemicals That Destroy Ozone

Chlorofluorocarbons are man-made chemicals that destroy the ozone. Chlorofluorocarbons are stable, non-flammable chemicals. They do not change or break down over time. Chlorofluorocarbons can stay in the stratosphere for as long as 100 years. During this time, they are broken down into hydrogen, fluoride, and chlorine by the sun's radiation. Each broken-down chlorine atom can destroy tens of thousands of ozone molecules. Other chemicals that are harmful to the ozone layer are nitrogen oxide (given off from cars and trucks) and halons (used in fire extinguishers).

# Ozone Layer Depletion    13

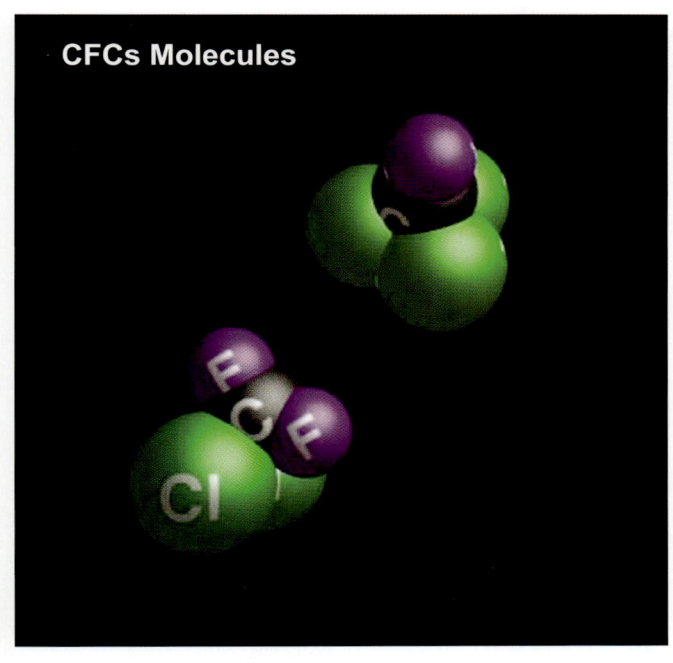

CFCs Molecules

## Dobson Unit

The *Dobson Unit* (DU) is the unit for measuring ozone in the atmosphere. One Dobson Unit is almost 27 million ozone molecules per square centimeter. The average amount of ozone in the ozone layer is about 300 DU.

## The Montreal Protocol

Discussions have been held to limit production and use of chemicals that harm ozone. The most important of these discussions was held in 1987 in Montreal, Canada. The industrialized nations created an agreement to decrease chlorofluorocarbons. The agreement is known as *The Montreal Protocol*.

## Ozone Hole

Every year a large "hole" forms in the ozone layer over the Antarctic. The hole is really a thinning of the ozone layer by almost 50%. The ozone hole was discovered in 1985.

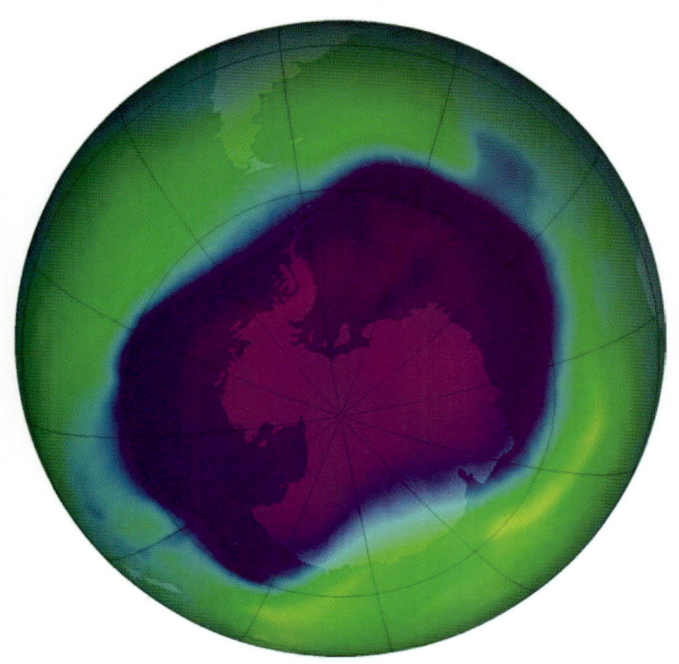

### Did you know?

An ozone hole has recently been found over the Arctic.

### How can we decrease ozone layer loss?

- Use energy-efficient appliances.
- Use less hot water.
- Choose a car that pollutes less.
- Buy air-conditioners or refrigerators that do not use chlorofluorocarbons.
- Do not buy fire extinguishers that contain halon.

# 14 THINK GREEN: Global Warming

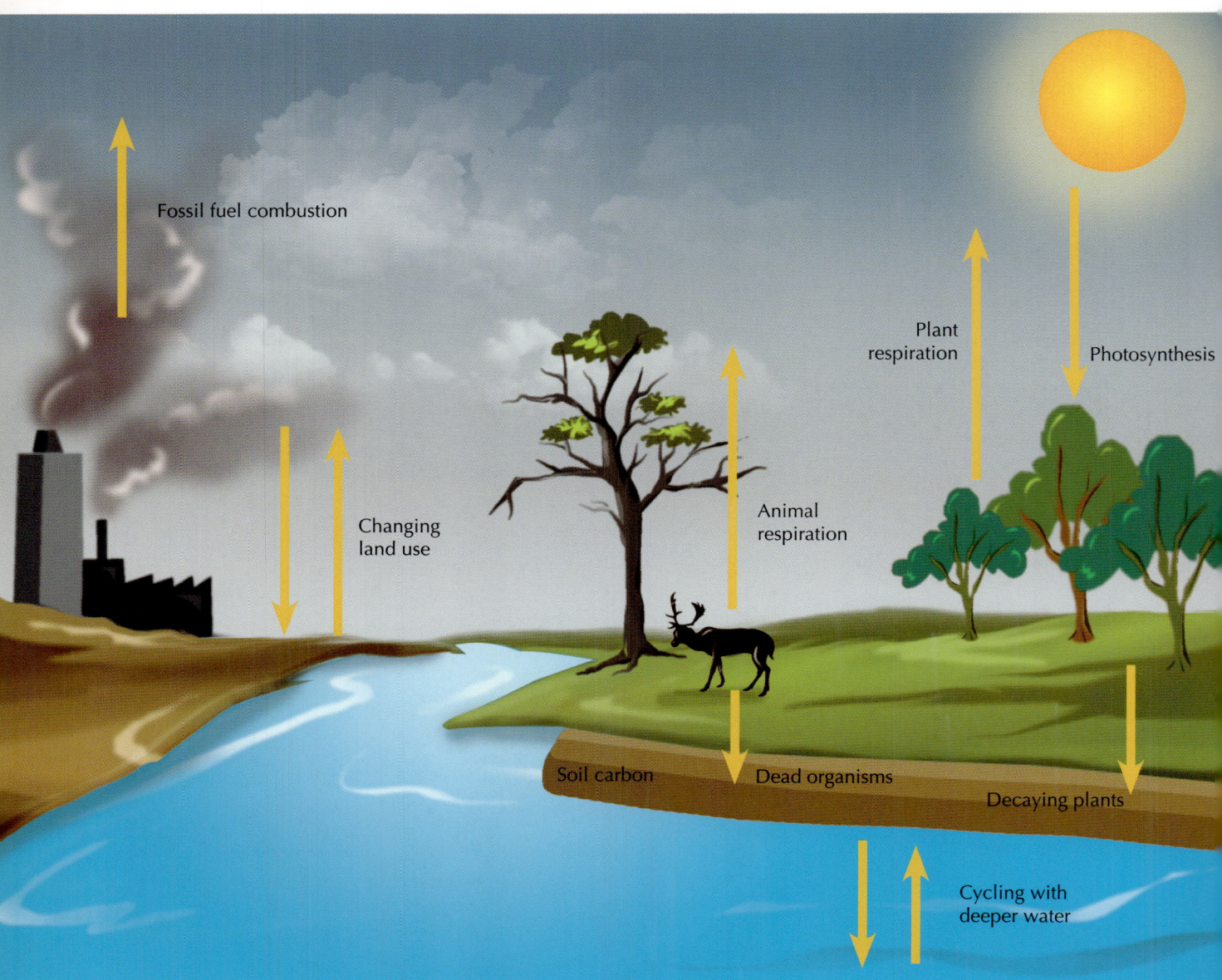

# The Carbon Cycle

Carbon is an element found almost everywhere on Earth. It is found in living and dead matter. Carbon is stored in soil, forests, and oceans. Fossil fuels are made up of carbon. Carbon moves in different forms through living beings, land, water, and air. This movement of carbon is known as the *carbon cycle*.

# The Carbon Cycle 15

## The Carbon Exchange

- Carbon is in the form of carbon dioxide in the air. It enters the living world through plants and algae. Plants use carbon atoms to make food through *photosynthesis*.
- Animals cannot absorb carbon dioxide from the air. They eat plants or other animals to get carbon.
- Carbon in living plants and animals is returned back to Earth when they die and their bodies break down, or *decompose*.
- Carbon dioxide also moves between the oceans and the atmosphere.
- Fossil fuel burning by industries and automobiles also releases carbon dioxide into the atmosphere.

### Did you know?

The amount of carbon dioxide in our atmosphere decreases during the warmer season. This is because there are more plants, and the rate of photosynthesis is high.

## Carbon Dioxide Cycle

The *carbon dioxide cycle* is one way carbon dioxide moves in the environment. Photosynthesis is the way plants absorb carbon dioxide to produce food using water. Plants release oxygen as a byproduct of photosynthesis. The oxygen is breathed by animals during *respiration*. Respiration is the reverse of photosynthesis. It takes in oxygen and gives off carbon dioxide.

## Where is carbon found in our planet?

**Carbon is found:**

- in living and dead organisms
- as organic matter in soil
- in gas form as carbon dioxide in the air
- in some rocks such as limestone
- as calcium carbonate shells in ocean life.

## Carbon Dioxide Emissions

Many human actions give off carbon dioxide into the atmosphere. Burning fossil fuels causes 70% to 75% of carbon dioxide emissions. Deforestation and automobiles add up to 25% of carbon dioxide in the atmosphere.

**THINK GREEN:** Global Warming

# The Kyoto Protocol

Most greenhouse gases are produced by industrialized nations. In 1997, more than 160 nations agreed to decrease their total emission of greenhouse gases. These nations met in the ancient Japanese capital of Kyoto and created an agreement. This agreement is known as the *Kyoto Protocol*.

# The Kyoto Protocol

## Signing the Kyoto Protocol

The Kyoto Protocol was signed by more than 140 countries. These countries promised to cut the amount of greenhouse gases they produced each year. The protocol came into effect on February 16, 2005, when Russia signed the protocol. The protocol could go into effect only if signed by nations accounting for at least 55% of greenhouse gas emissions. The United States has since withdrawn from the Kyoto Protocol.

## The Agreements in Kyoto Protocol

The Kyoto Protocol commits 38 industrialized countries to limit their greenhouse gas emissions. The goals of the protocol are as follows:

- Individual targets have been set for each country based on its pollution levels.
- Targets have been set for 2008 through 2012.
- Cuts in carbon dioxide, methane, and nitrous oxide will be measured against a base year of 1990.
- Growing, developing countries such as China and India are outside the framework.

## Countries That Can Increase Emissions

- Iceland by 10%
- Australia by 8%
- Norway by 1%

## Percentage of Greenhouse Gas Reduction

- Switzerland and the European Union by 8%
- The Central and East European countries (including Bulgaria, Czech Republic, and others) by 8%
- The United States by 7%
- Canada, Hungary, Japan, and Poland by 6%

### Did you know?

Japan is Asia's third largest carbon dioxide producer. It is also the fifth biggest polluter in the world.

**THINK GREEN:** Global Warming

# Climate Change

*Climate* is the average weather conditions in an area over several years. Climate change is the *variation*, or change, in the climate of a place. It includes changes in temperature and precipitation. Climate change may occur for many reasons. Climate change is taking place around the planet. It has caused rising sea levels, melting glaciers, higher floods, and stronger storms.

## Causes of Climate Change

### Continental Drift

Around 200 million years ago, all the continents were joined together as a giant landmass or supercontinent. This giant landmass began to drift apart. Over many millions of years, it formed Earth's continents. This *continental drift* changed the physical features and position of land and oceans. Continental drift affected the climate of all places on earth.

### Volcanoes

Volcanoes erupt and release large amounts of water vapor, sulfur dioxide, dust, and ash into the atmosphere. These particles block some of the sun's rays. This cools Earth. The sulfur dioxide also combines with water to form tiny droplets of sulfuric acid. This reflects sunlight and keeps some of the energy from reaching the ground.

Climate Change  19

### Earth's Tilt
Changes in Earth's tilt affect how severe the weather is. More tilt means warmer summers and colder winters. With less tilt, summers will be cooler and winters warmer.

### Ocean Currents
Ocean currents absorb the sun's rays and move large amounts of heat across Earth. Heat escapes from the oceans in the form of water vapor. Water vapor helps form clouds and has a cooling effect.

> **Did you know?**
> Rice is the world's most important grain crop. Recent studies have shown that rice yields fall by 10% for each 1°F of global warming.

### Effects of Climate Change
- Melting glaciers and the spread of warmer seawater cause the sea level to rise.
- Heavy rainfall causes floods.
- Higher temperatures cause some areas to dry out.
- Warm temperature harms some kinds of animals.
- Changes in temperature and rainfall affect crop yields.

### Stopping Climate Change
- Use *renewable* sources of energy.
- Use energy efficiently. This means wasting as little as possible.
- Prevent deforestation.

### Human Causes of Climate Change
- Power plants and cars burn fossil fuels and give off greenhouse gases and pollutants.
- Plastic waste stays in the environment and causes damage.
- Increased use of paper and paper products leads to deforestation as more trees are cut down.

# Severe Weather

Severe weather is any harsh or strong weather event that is a threat to human life and property. Severe thunderstorms, tornadoes, hurricanes, heat waves, snowstorms, floods, and hail are all types of severe weather. Many scientists think that global warming has led to an increase in severe weather.

## Thunderstorms

Thunderstorms are a type of severe storm. These storms cause the most damage. Thunderstorms produce tornadoes, lightning, high winds, heavy rain, and hail.

## Flash Floods

Flash floods need heavy rain to occur. When thunderstorms move slowly, flash floods may happen. Flash floods that come with thunderstorms cause hundreds of deaths every year.

## Tornadoes

Tornadoes, or twisters, are funnel-shaped rotating columns of air. They form at the bottom of a thundercloud and touch the ground. Tornadoes are rated according to a scale known as the F-scale or Fujita scale. The scale was developed by Dr. Theodore Fujita in 1971. The F-scale contains six categories, or groups. They range from F0 for the weakest to F5 for the strongest tornadoes. The scale rates tornadoes according to wind speed and damage.

# Severe Weather Safety Rules

## Tornadoes
- When tornadoes occur, go to the lowest floor of the house. If there is no basement, go to an inside room.
- Protect yourself from flying objects by wrapping blankets or coats around your body.
- Avoid sitting in cars or mobile homes during tornadoes.
- Try to be in a designated tornado shelter.
- Cover your head with your hands and lie down flat in the nearest ditch or low area.

## Lightning
- Do not use the telephone or other electrical appliances.
- Avoid taking a shower or bath.
- If you are in the water, immediately get out of the water and move to a safe area.
- If you are in a wooded area, move under a thick growth of small trees.
- Do not lie flat on the ground.

## Flash Floods
- Avoid areas that are likely to flood while driving.
- If your vehicle stalls in floodwaters, get out of it immediately and leave it.
- Try to climb to higher ground.

## River Floods

### Before
- Keep a first aid kit ready.
- Try to know about your flood risk.
- Have enough fuel in your vehicle.
- Store as much drinking water as possible.
- Store some food products.
- Keep a radio handy.

### Did you know?
Tornadoes in the northern hemisphere spin counter-clockwise. In the southern hemisphere they spin clockwise.

### During and After
- Avoid driving on flooded areas.
- Do not let children play near high water or storm drains.
- Boil water before drinking it.
- Do not eat food that has been exposed to flood waters.
- Make sure electrical appliances are dry before using them.
- Use flashlights instead of lanterns or matches inside a building.

| Original Fujita Scale | | Operational Fujita Scale (mph, 3-second gust) |
|---|---|---|
| F0 Gale Tornado 40-72 mph | Damage to chimneys and signboards. Broken tree branches and uprooted shallow-rooted trees. | 65-85 |
| F1 Moderate Tornado 73-112 mph | Hurricane-force winds begin in the lower limit. Surface of roofs peeled off. Mobile homes and automobiles pushed over. | 86-110 |
| F2 Significant Tornado 113-157 mph | Roofs torn off frame houses. Mobile homes demolished. Boxcars pushed over, large trees uprooted. Generates light-object missiles. | 111-135 |
| F3 Severe Tornado 158-206 mph | Severe damage. Roofs and walls torn off well-constructed homes. Overturned trains. Most forest trees uprooted. Heavy cars thrown. | 136-165 |
| F4 Devastating Tornado 207-260 mph | Well-constructed homes destroyed. Structures with weak foundations blown off. Cars thrown and large missiles generated. | 166-200 |
| F5 Incredible Tornado 261-318 mph | Phenomenal damage. Strong frame homes disintegrate or are lifted off foundations and carried considerable distance. | over 200 |

**THINK GREEN:** Global Warming

# Heat Waves

A heat wave is an unusually large rise in temperature in an area. Temperatures during a heat wave rise above 90°F or more. They can last from a few days to several weeks.

## Are heat waves dangerous?

Heat waves are not as dangerous as other extreme weather events such as floods, tornadoes, and earthquakes. However, heat waves have caused many deaths.

## Heat-Related Health Problems

- Severe sunburn: Sunburn damages the skin and decreases the skin's ability to release heat.
- Heat cramps: Heat cramps and spasms are caused by overuse of muscles. Too much water is lost from the body through heavy sweating.
- Heat exhaustion: Heat exhaustion leads to weakness, heavy sweating, and clammy skin. It causes a weak pulse, vomiting, and fainting.
- Heat stroke: Heat stroke is also known as sunstroke. The body cannot control its own temperature. Increased body temperature causes brain damage and even death if not treated immediately.

## Heat Index

Heat index gives a measure of how hot it really feels. To figure out the heat index, the amount of water in the air is added to the air temperature. Heat index rises by 15°F when exposed to full sunshine.

> ### Did you know?
> In the second half of the 21st century, heat waves in Europe and North America may become more common and extreme.

| Relationship Between Heat Index and Heat Disorder | |
|---|---|
| 130° F or higher | Heatstroke/sunstroke highly likely with continued exposure |
| 105-130° F | Sunstroke, heat cramps, or heat exhaustion likely with prolonged exposure |
| 90-105° F | Sunstroke, heat cramps, or heat exhaustion possible with prolonged exposure and/or physical activity |
| 80-90° F | Fatigue possible with prolonged exposure and/or physical activity |

## What to Do During a Heat Wave

- Drink plenty of water even when not feeling thirsty. Water keeps the body cool.
- Do not drink liquids with alcohol or caffeine.
- Avoid salt.
- Eat small and regular meals.
- Wear lightweight and light-colored clothes. Light colors reflect sunlight.
- Stay inside the house in air-conditioned rooms.
- Slow down and avoid hard exercise.

## How to Treat Heat Emergencies

- **Heat cramps:** Take the person to a cooler place. Let him or her rest. Stretch the affected muscles lightly. Every 15 minutes, give the person a half glass of cool water.
- **Heat exhaustion:** Take the person to a cooler place. Remove tight clothes and give wet and cool clothes. Give cool water to drink.
- **Heat stroke:** Take the person to a cooler place. Give the person a cool bath or wrap wet sheets and fan. Keep the person lying down. Do not give anything to drink or eat if the person refuses water or feels sick.

**THINK GREEN:** Global Warming

# Earlier Spring

Every season occurs in a certain range of temperatures. Change in temperatures leads to changing seasons. In the last 100 years, global warming has warmed Earth by about 1°F. During this time, the average low temperature of Earth has increased. This has made the warm season longer. In the northern hemisphere, spring now comes about 1.2 days earlier every 10 years. Earlier spring affects the life cycle of plants and animals. This can harm the entire ecosystem.

## Earlier Spring and Thawing of Frozen Soil

For thousands of years, the frozen soil in the *tundra* and *boreal forests* have absorbed large amounts of carbon. Earlier warm temperature causes early *thawing*, or melting, of the frozen soil. This releases the stored carbon into the atmosphere. The released carbon reacts with oxygen in the air to make carbon dioxide. This increases greenhouse gases in the atmosphere, which adds to global warming.

## Earlier Spring and Animal Adaptation

Spring is the breeding time for plants and animals. Earlier spring has shortened the hibernation period of mammals. *Hibernation* is sleeping through the winter. Earlier spring has also made birds lay eggs earlier. Also, many kinds of fish are moving northward in search of cooler waters.

### Did you know?
Loggerhead sea turtles of Florida's Atlantic coast now lay eggs 10 days earlier than they did 15 years ago.

## Earlier Spring and Blooming

Spring is the blooming season for flowers and trees. Plants have now begun to bloom one or two days earlier every 10 years. Apples and grapes are blooming six to eight days earlier than they did in 1965.

## Earlier Spring and Bird Migration

Birds migrate when weather changes, when food is harder to find, and for breeding. In summers, the birds move to cooler places. Now with the early arrival of spring, birds are migrating earlier than usual.

**THINK GREEN:** Global Warming

# Heavy Rain

Water *evaporates* to form a gas, and *condenses* to form liquid droplets. Droplets in the sky form clouds. Rain is water falling in drops from clouds. Rain is an important part of the water cycle on Earth. A warmer atmosphere contains more water vapor. That leads to higher rainfall.

### Heavy Rain in Northern Hemisphere

Global warming is causing heavy rain across parts of the northern hemisphere. New studies by scientists have shown that this increase of heavy rain is caused by greenhouse gases that come from human actions.

### Water Can Be Dangerous

Even though it is important for life on Earth, water can also be dangerous. Heavy rain occurs when the precipitation rate is greater than 0.3 inches per hour. A *monsoon* is a seasonal wind or air mass caused by heating. Monsoons bring very heavy rain. The largest monsoon is found in the Indian Ocean and southern Asia.

Heavy Rain 27

### During Heavy Rain
- Avoid driving during heavy rain. About 80% of flood deaths occur in vehicles.
- Do not walk through water, streams, or rivers when they are in full flow.
- Leave your homes and offices after turning off the utilities.

### After Heavy Rain
- Check the damage to houses and buildings.
- Use a flashlight in darkness instead of candles.
- Throw away food, even canned foods, that have been exposed to floodwater.

## Flood
Floods are caused by heavy rain that continues for a long time. Heavy rain pours lots of water into rivers and lakes. The water overflows and covers land.

## Before Heavy Rain
- Move important things to places at a higher level.
- Bring objects that may be damaged inside the house.
- Keep drinking water with you.

### Did you know?
Each year, about 80 people are killed and nearly 300 are hurt by lightning in the United States.

**THINK GREEN:** Global Warming

# Heat Harms the Atmosphere

The *atmosphere* is the thin layer of gases that surrounds Earth. It protects Earth from the harmful rays of the sun. The atmosphere is about 300 miles thick and is divided into several layers. The temperatures in these layers are different.

## Composition of the Atmosphere

The atmosphere is mostly made up of two gases: nitrogen and oxygen. Nitrogen is 78% and oxygen is about 21% of the atmosphere. It also contains argon, carbon dioxide, water vapor, and small amounts of other gases known as *trace gases*.

## Keeping Earth Warm

Earth is kept warm by the atmosphere. The gases in the atmosphere allow sunlight to pass through but stop most of the energy from escaping into space. This effect is known as the natural greenhouse effect. This is what keeps Earth warm.

Heat Harms the Atmosphere  29

## Greenhouse Gases

Greenhouse gases absorb or keep radiation from escaping into space. Water vapor, carbon dioxide, methane, nitrous oxide, and chlorofluorocarbons are important greenhouse gases. In the last century, more greenhouse gases have been added to the air. The main reason for this increase in greenhouse gases is the burning of fossil fuels.

### Did you know?
The years 1998 and 2006 were among the warmest of the last hundred years.

## Effects of Increasing Greenhouse Gases

The increased emission of greenhouse gases has led to global warming. The major effects of a temperature rise are:

- Very large changes in climate
- Extinction of many species of plants and animals
- Rising sea levels
- More extreme weather
- Spread of diseases
- Changes in the types of crops being grown in different parts of the world

**THINK GREEN:** Global Warming

# Heat Increases Hurricane Strength

Hurricanes are very strong tropical storms. They start in the oceans close to the equator. Hurricanes come with heavy rainfall and winds blowing at speeds of 74 mph or more. Hurricanes cause a lot of property damage and kill many people. Global warming does not create hurricanes, but it makes them stronger.

### Peak Season for Hurricanes
Hurricanes have seasonal patterns. The peak season for hurricanes in the northern hemisphere is June to November. In the southern hemisphere, most hurricanes occur from January to March.

### Global Warming Has Increased the Strength of Hurricanes
Scientists believe that hurricanes have become stronger over the past 30 years. Global warming has led to a rise in the surface temperature of tropical ocean water. This has caused a 50% increase in the wind speeds of hurricanes. It has also made hurricanes last longer.

### Parts of a Hurricane
**Eye**
The *eye* is the center of a hurricane. Spiral bands of the storm rotate around the eye. The eye does not have any clouds. It does have some wind and rain. It is much calmer than the rest of the hurricane.

**Eye Wall**
The *eye wall* is a circle of clouds that surrounds the eye of the hurricane. It looks like a bright cloud ring right around the eye. It is the strongest area of the hurricane.

**Spiral Bands**
*Spiral bands* are clouds that surround the eye wall. They form the largest area of a hurricane. They produce rain and fierce winds. Most flooding occurs because of the rain from spiral bands.

## Stages of a Hurricane

1. **Tropical disturbance**: Hurricanes begin as *tropical disturbances*. In this stage, thunderclouds start to group together. Wind speeds are less than 25 miles per hour.
2. **Tropical depression**: A tropical disturbance becomes a *tropical depression* when the winds begin to swirl in a *cyclonic* (spiral or circular) pattern. Winds near the center are constantly between 23 and 39 miles per hour.
3. **Tropical storm**: A depression becomes a *tropical storm* when wind speeds stay at 39 to 73 miles per hour. These storms are given names.
4. **Hurricane**: A tropical storm becomes a hurricane if the surface pressures continue to drop and winds reach 74 miles per hour. A hurricane has a well-defined eye with spiral rain bands rotating around it.

## Hurricane Ratings

Hurricanes are rated according to a scale known as the *Saffir-Simpson scale*. The scale contains five categories. It rates a hurricane according to wind speed, storm surge, and the amount of damage that it could produce.

### Did you know?
Hurricane Katrina in August 2005 was the most destructive natural disaster in the history of the United States. It killed about 1,200 people and damaged property worth $75 billion.

## Damages Caused by Hurricanes

There are three main kinds of damage caused by hurricanes:

1. **Wind damage**: Hurricane winds can destroy mobile homes and small homes. Some items, such as signs and roofing material, become flying missiles in hurricanes.
2. **Storm surge damage**: The onshore rush of high ocean waves caused by a hurricane is known as a *storm surge*. The surge of water that sweeps across the coastline causes severe flooding. It is the greatest threat to human life and property.
3. **Flood damage**: Very heavy rains can cause deadly and destructive floods.

**THINK GREEN:** Global Warming

# Arctic Ice Melting

The Arctic is a huge area near the North Pole. It is covered in ice and snow all year. The Arctic is very important in the global weather system. The snow and ice form a cooling layer that reflects energy from the sun. This helps to keep Earth cooler. With the rise in average temperature, the Arctic ice caps are melting.

### Reasons for Arctic Ice Melting
Scientists have found two main reasons for Arctic ice melting. These are rising air temperature and warmer seawater. The rising air temperature melts the ice from above. The warm seawater attacks the ice from below.

### The Arctic Now
The Arctic has changed in the last 100 years. It is being affected by changing global climate. The ice cover has decreased. The shrinking ice cover has affected many Arctic animals.

## Effects of Arctic Ice Melting

- Glaciers are shrinking and sea levels are rising around the world.
- Native people, plants, and wildlife have been affected by the melting of Arctic ice.
- The number of storms in coastal areas have increased. This causes soil to wash away.
- The Arctic ice melting raises the levels of ultraviolet radiation.

## Effects of Ward Hunt Ice Shelf Break-up

The break-up of ice shelves creates icebergs that float in the sea. These icebergs are a danger to ships. They also affect oil rigs. The release of freshwater may also affect the environment.

## Ward Hunt Ice Shelf Break-up

The Ward Hunt Ice Shelf is the Arctic's largest ice shelf. It has been in place for about 3,000 years. In the summer of 2002, graduate student Derek Mueller discovered that the Ward Hunt Ice Shelf was breaking up. Scientists believe that the break-up of the shelf was due to the increase in the average temperature of the area.

## The Effects of Ice Shelf Loss

The loss of sea ice can have a big effect on global climate. The loss of ice shelves may affect wildlife and human life. Sea ice reflects most of the sun's energy. When the ice melts, it becomes seawater. This seawater absorbs most of the sun's energy. The melting of ice shelves means that more radiation will be absorbed, making Earth hotter. This adds to global warming.

### Did you know?

About 81% of Greenland is covered with ice. This ice contains enough water to raise the global sea level by about 21 feet.

# Worldwide Pollution

Pollution is any form of *contamination*, or mixing with harmful materials. Land, water, and air can be polluted. Pollution is harmful to human health and causes damage to the environment. Pollution is divided into four types: air pollution, water pollution, land pollution, and noise pollution.

## Air Pollution

Air pollution is caused by chemicals in the air. Industries and cars that burn fossil fuels are two main sources of air pollution. Other sources of air pollution are natural radioactivity, volcanic eruptions, and forest fires.

## Water Pollution

Water is polluted by farming, homes, or industries. Sewage and fertilizers used in farming contain nitrates and phosphates that make some plants and algae grow more. The extra growth decreases the amount of oxygen in the water. This means that the fish and other animals that need oxygen can't breathe well.

### Did you know?
Nearly 40% of rivers and 46% of lakes in the United States are too polluted for swimming and fishing.

## Land Pollution

Poor use of land resources causes land pollution. This causes soil erosion and *salinization*, or contamination with salt. Harmful chemicals, radioactive waste, sewage sludge, and other types of harmful waste find their way to landfill sites.

## Radioactive Pollution

Radioactive pollution comes from nuclear reactions. Nuclear power plants and nuclear weapons create radioactive waste. The transportation and disposal of nuclear waste cause radioactive pollution.

### Causes of Land Pollution
- Agriculture
- Mining and quarrying
- Sewage sludge
- Household wastes
- Industrial wastes

## Noise Pollution

*Noise pollution* is a serious problem in most cities. Most modern towns and cities are noisy. Noise can affect our hearing and cause other health problems.

### Sources of Noise Pollution
- Road traffic
- Air traffic
- Rail traffic
- Neighborhood and household noise
- Industrial noise

### What can you do to prevent pollution?
- Recycle as much waste as you can.
- Avoid using things like paper plates and cups that become waste.
- Walk or use a bicycle.
- Make compost of waste food.

**THINK GREEN:** Global Warming

# Air Pollution

Air pollution is the contamination of air. These harmful pollutants often enter our bodies when we breathe. They can cause many serious diseases. The main sources of air pollution are factories, power plants, cars, buses, trucks, and wildfires.

## Major Air Pollutants

### Carbon Monoxide (CO)
Carbon monoxide is a gas that has no color and no smell. It is produced when common fuels are not burned completely. It also comes from the burning of products such as cigarettes.

### Carbon Dioxide ($CO_2$)
Carbon dioxide is the most important greenhouse gas. It is produced mostly by burning coal, oil, and natural gas.

### Chlorofluorocarbons (CFCs)
Chlorofluorocarbons are man-made chemicals containing carbon, chlorine, fluorine, and sometimes hydrogen. These chemicals are commonly used in air conditioners and refrigerators.

### Lead
Lead is in gas, diesel, lead batteries, paints, hair dye, and other products.

### Ground Level Ozone
Ground level ozone is given off by vehicles and industries.

### Nitrogen Oxides (NOx)
Nitrogen oxides are produced from burning fuels including gas, diesel, and coal.

### Suspended Particulate Matter (SPM)
*Suspended particulate matter* (SPM) is made up of tiny particles that remain hanging in the air. SPM occurs in the form of smoke, dust, and vapor.

### Respirable Particulate Matter (RSPM)
*Respirable particulate matter* (RSPM) are tiny pollutant particles of liquids, soot, dust, smoke particles, and acids from aerosols. These particles are small enough to be breathed into our lungs.

### Sulfur Dioxide ($SO_2$)
Sulfur dioxide is produced by the burning of fossil fuels and by some industries.

## Health Hazards

Air pollution can cause problems for humans, plants, and animals. Some of the main health risks in humans are lung, heart and brain diseases.

The following table shows different air pollutants and the health risks they cause:

| Name of pollutant | Health Risks |
|---|---|
| RSPM | Respiratory illness, including chronic bronchitis and asthma; heart disease |
| $SO_2$ | Heart diseases; respiratory problems, cancer, eye burning, headache |
| $NO_2$ | Lung problems, viral infection, chest tightness |
| SPM | Lung diseases, asthma, cancer |
| Benzene | Immune system disorders, increased risk of cancer, asthma, anemia |
| Ozone | Reduced lung function, chest pains, coughing, irritation of eyes and nose |
| CO | Cherry lips, unconsciousness, death by *asphyxiation* (not getting enough clean air) |
| Lead | Decreased hemoglobin in blood or anemia, damage to the nervous system |

## Cause Less Air Pollution

- Walk, ride your bicycle, or use public transportation whenever possible.
- Form a car pool.
- Use fewer aerosols.
- Plant trees.
- Switch off all electrical appliances when not in use.

## Smog

*Smog* is haze (like fog) caused when chemicals that pollute the air react in sunlight. They combine with water vapor and dust. Low wind speeds cause smoke and fog to sit in one place. The main parts of smog are ground-level ozone, nitrogen oxides (NOx), volatile organic compounds (VOCs), sulfur dioxide, and particulate matter.

## Acid Rain

Rainwater is *acidified* (made acid) by some air pollutants. Sulfur dioxide and nitrogen oxides are the main causes. *Acid rain* damages plants, poisons the soil, and harms animals and fish.

## Indoor Air Pollution

Smoking, cooking, heating appliances, and paints are some of the main sources of indoor air pollution.

## Sources of Air Pollution

- Power plants, manufacturing plants, and waste incinerators
- Motor vehicles, aircraft, and ships
- Burning wood, fireplaces, stoves, and furnaces
- Chemicals and dust
- Vapors from paint, hair spray, and aerosol sprays
- Waste in landfills

### Did you know?

Americans have decreased air pollutants by more than 50 million tons since 1970.

**THINK GREEN:** Global Warming

# Driving

Automobiles add to air pollution. Most automobiles use gasoline as fuel. The burning of gasoline gives off three main pollutants: hydrocarbons, carbon monoxide, and nitrogen oxides. These pollutants increase the greenhouse effect. They also cause smog. They harm human health.

### Alternative Fuel Vehicles
Alternative fuel vehicles run on fuels that are not made from petroleum (oil).

# Driving

## Alternative Fuels

| Fuel | Source/type |
|---|---|
| Ethanol | Produced from corn and other crops |
| Biodiesel | Produced by mixing vegetable oils or animal fats with alcohol |
| Compressed natural gas (CNG) | Fossil fuel |
| Liquefied petroleum gas (LPG) | Fossil fuel |
| Hydrogen | Produced by nuclear power, by burning fossil fuels, or by renewable resources, such as water power |

## Hybrid-electric Vehicles

*Hybrid-electric vehicles* are powered by both an internal combustion engine (gas) and an electric motor. Hybrid cars use less fuel and reduce tailpipe emissions. Therefore they produce less pollution.

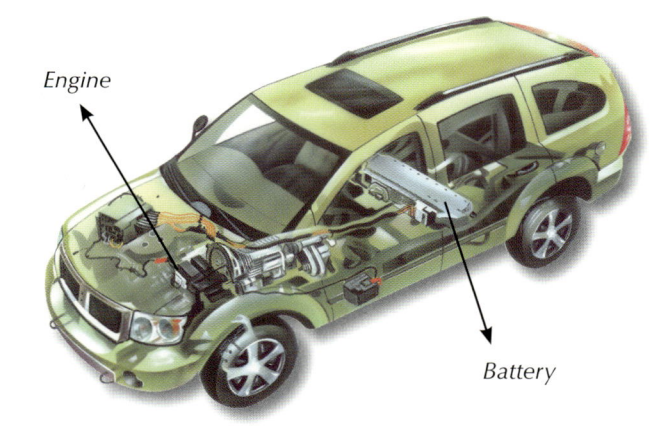

*Engine*

*Battery*

## Electric Vehicles

*Electric vehicles* are powered by electric motors and rechargeable batteries. They are energy efficient and environment friendly. Electric vehicles do not give off tailpipe emissions.

## Tips for Green Driving

All drivers can help reduce air pollution by changing their driving habits:

- Avoid driving at high speed.
- Buy vehicles that cause less pollution.
- Keep tires properly inflated.
- Maintain your vehicle's air conditioning system and regularly check for leaks.

## Top Five Global-Warming Polluters

1. United States
2. China
3. Russia
4. Japan
5. United Kingdom

### Did you know?

One gallon of burned gasoline puts 28 pounds of carbon dioxide into the atmosphere. An average car gives off about 63 tons of carbon dioxide over its lifetime.

**THINK GREEN:** Global Warming

# Public Transportation

Public transportation systems are often provided by the government. They may include buses, trams, trains, cable cars, monorails, and ferryboats. Using public transportation is better for the environment. It can help decrease carbon dioxide emissions and save several million gallons of gasoline.

### Benefits of Public Transportation
- Saves energy by reducing the amount of fossil fuel burned.
- Decreases emission of harmful gases. This improves air quality.
- Creates jobs and helps the economy.
- Causes less traffic and saves time.

## Air Transportation

Air transportation connects different places across the world. It is usually the fastest way to travel. Aircraft release harmful gases and particles.

## Heavy Rail Transit

Heavy rail (also known as metro, subway, rapid transit, or rapid rail) is an electrified, high-speed train. Some heavy rail transit systems are the Chicago "L", the New York City Subway, the London Underground, and the Delhi Metro.

### Did you know?
Each year, using public transportation saves nearly 1.4 billion gallons of gas in the United States.

## Light Rail

Light rail (also known as streetcar, tramway, or trolley) is similar to heavy rail transit. It is used for shorter distances. It can carry fewer passengers.

## Trolleybus

Trolleybuses (also called trackless buses or trackless trams) are electric buses that are powered by overhead wires.

## Bus

Buses are the most common means of public transit. Buses, however, are not environmentally friendly. They run on gasoline or diesel which adds to air pollution. Hybrid buses cause less pollution and are better for the environment. Some buses also run on alternative fuels like compressed natural gas (CNG) and biodiesel.

**THINK GREEN:** Global Warming

# Permafrost

*Permafrost* is a layer of soil that is frozen all the time. Permafrost occurs at high altitudes and near the North and South Poles. Its thickness ranges from less than three feet to more than 3,000 feet. Permafrost preserves organic materials. It also slows water movement and plant growth. Permafrost covers nearly 25% of Earth's land surface.

## Global Warming and Permafrost

Permafrost contains large amounts of carbon in the form of grass and animal bones. This material has been trapped inside the permafrost for thousands of years. Global warming could thaw or melt this permafrost. Then the carbon would be released into the atmosphere as a greenhouse gas.

## Permafrost Zones

### Continuous Permafrost Zone

The *continuous permafrost zone* is an unbroken layer of frozen soil. This zone occurs in the Arctic area. Greenland is within this zone.

### Discontinuous Permafrost Zone

The *discontinuous permafrost zone* is an irregular area that has scattered pockets of unfrozen ground. Fairbanks, Alaska, is within this zone. The discontinuous permafrost zone is divided into two sub-zones: the *widespread permafrost zone* and the *sporadic permafrost zone*.

## Permafrost 43

### Thawing Permafrost in the Arctic

The Arctic *region* (area) has large zones of permafrost. The Arctic stores about 14% of the carbon that is found in Earth's soil. Thawing of the permafrost will release carbon and methane. This would increase global warming.

### Thawing Permafrost in Alaska

About 85% of Alaska has permafrost. Global warming has led to thawing of the permafrost. This has caused a drop in water and plants. It also causes soil to wash into rivers. The soil clogs the rivers and makes them shallow.

### Effects of Melting Permafrost

- Loss of soil quality due to erosion
- Release of carbon and methane that add to global warming
- Unstable ground for buildings and roads
- Climate change that could cause economic problems in an area

### Did you know?

Almost one-half of the world's permafrost is in Russia and Siberia and one-third is in Canada.

**THINK GREEN:** Global Warming

# Green Homes

Green homes are buildings that are better for the environment, or *eco-friendly*. Green homes use less energy and water. They are made from materials that do not harm the environment. They also protect human health. Green homes have cleaner indoor air. They use renewable resources such as sunlight, rainwater, and biomass.

Green Homes  **45**

## Why do we build green homes?

Green homes help the environment and the economy. Green homes are energy-efficient. They have lower electricity, water, and gas bills. Green homes produce less waste and use recycled materials in their construction.

### Did you know?

In the United States, buildings are responsible for 39% of total energy use, 12% of the total water use, and 68% of total electricity use.

## Energy Star

Energy Star is a U.S. government program that certifies and labels products that are energy-efficient for use in homes. Energy Star helps to cut energy bills in homes. It rates the efficiency of appliances, like air conditioners, fans, refrigerators, and computers.

## Improving Energy Efficiency at Home

- Plug air leaks
- Use less water
- Use energy-efficient windows
- Use energy-efficient lighting
- Use energy-efficient appliances

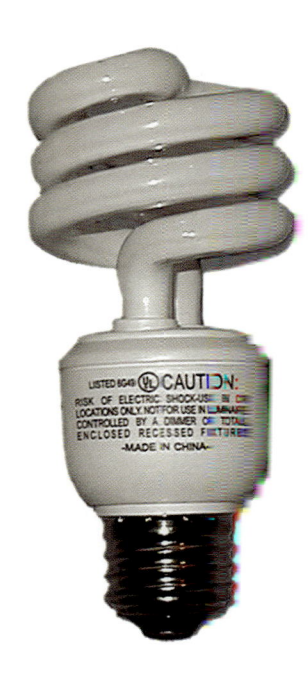

**THINK GREEN:** Global Warming

# Collaborative Housing

Collaborative housing, or cohousing, is a type of planned community. It combines the privacy of individual homes with the advantages of living in a community. Residents, often known as cohousers, plan, design, and maintain the community themselves.

# Collaborative Housing

## Cohousing Saves Valuable Resources
The idea behind cohousing is to save resources. Most cohousing communities are environment-friendly. Cohousing saves heating and maintenance costs. It also decreases energy use.

### Did you know?
A kibbutz is a collaborative community in Israel. The first kibbutz was built on the bank of the Jordan River in 1909.

## Characteristics of Collaborative Housing
**All-round Participation**: Usually, cohousers participate in all the parts of building and living in the community. This helps them to meet all their basic and long-term needs.

**Community Design**: The buildings are designed to create a sense of community among cohousers. Typically, all houses face each other across a pathway or courtyard. A common house is built in a central place. It serves as a community center. Cars are parked on the outer edge to keep the housing area free of pollution and of cars.

**Community Facilities**: The common areas include a dining area, sitting area, kitchen, play area, library, laundry, exercise room, lawns and gardens, and guest rooms.

**Managing the Community**: A cohousing community is managed and developed by the cohousers themselves. Cohousers meet regularly to solve the problems of the community. The residents develop the rules for the community. Cohousers also share all work within the community.

## Cohousing and the Environment
Cohousing helps reduce energy use by making common meals and using a common laundry. Cohousing communities have open green space for play and to help the environment. There are community gardens where people grow vegetables. Residents share cars for trips to the store and other places.

## Benefits of Cohousing
- Ideal balance between privacy and community
- Social interaction and support
- Low living cost
- Savings on meals
- Reasonably priced homes
- Safe environment
- Better childcare
- Sharing of resources

**THINK GREEN:** Global Warming

# Deforestation and Altering the Carbon Balance

Forests cover about a third of Earth's total land area. They are needed for all life on earth. Human actions, pollution, and acid rain are slowly destroying forests. Most of this *deforestation* takes place to provide land for farming, housing, or industrial uses.

### Carbon Reservoir

Forests are the most efficient ecosystem on Earth. They absorb carbon dioxide from the atmosphere. They store it as carbon in wood, plants, and soil. Most of the carbon dioxide in the atmosphere comes from burning fossil fuels. Earth's forests act as a *reservoir*, or collection site, for carbon dioxide.

### Forests and Greenhouse Gases

Deforestation releases huge amounts of carbon dioxide into the atmosphere. About 17% of the carbon dioxide in the atmosphere comes from deforestation.

# Deforestation and Altering the Carbon Balance

## Earth's Forest Cover
Most of Earth's forests have disappeared. More than 80% of Earth's natural forests may have already been destroyed. Logging, land clearing, and forest fires are responsible for this loss of forests.

## How Deforestation Happens
Deforestation is brought about by the following:
- Cutting forests and woodlands for farmland or development
- Cattle ranching
- Commercial logging
- Cutting trees for firewood

## Forests and Rainfall
Forests absorb large amounts of sunlight. Forests use sunlight for photosynthesis. They absorb almost 90% of all light. Photosynthesis is the process by which plants convert carbon dioxide. The energy absorbed by forests helps to form *convection currents* in the air. Convection currents are caused by the rising and falling of hot air. They help to increase the amount of rainfall in forests. On the other hand, deforested areas absorb only about 80% of sunlight. Deforested areas therefore become drier with time. They may turn *arid*, or dry like a desert.

## The Effects of Deforestation
- Destruction of environments that are one of a kind
- Extinction (dying out) of animals and plants
- Soil becomes dry and cracked because of exposure to the sun
- Loss of shaded area leads to hotter and colder temperatures
- Loss of plants used in medicine
- No recycling of water
- It is hard for rain to soak into the soil, so flooding may occur.

## What can we do?
- Replant trees in deforested areas.
- Limit activities like logging and over-grazing.
- Buy recycled products.
- Create public awareness.

### Did you know?
Forests are home to about 70% of all animals and plants.

**THINK GREEN:** Global Warming

# Fossil Fuels

Fossil fuels are energy resources found in Earth's crust. They formed many hundreds of millions of years ago from the remains of plants and animals. Coal, oil, and natural gas are the three main forms of fossil fuels. They are nonrenewable energy sources. Fossil fuels provide most of the energy used to make electricity, to heat and cool our homes, and to power motor vehicles and airplanes. More fossil fuels are used because the world population is growing. The burning of fossil fuels increases global warming.

### Benefits of Fossil Fuels

Fossil fuels are an important source of energy. They are useful as fuel because of their low cost and easy availability. Coal is the most common fossil fuel. It is found worldwide.

## Coal

Coal is a hard, rock-like substance. It is formed from the remains of dead plants over millions of years. It is mostly carbon. Coal is taken out of Earth by mining. There are three main types of coal: anthracite, bituminous, and lignite. Each type contains different amounts of carbon. Coal is used in industries and homes.

### Did you know?

Coal is a cheaper fuel than most others. Coal can provide energy at $1 to $2 per million Btu. Oil and natural gas cost $6 to $12 per million Btu.

## Natural Gas

Natural gas is another form of natural fossil fuel. Natural gas is mostly methane. Natural gas gives off fewer harmful chemicals than oil or coal. It is used for cooking as well as to produce electricity. Natural gas can be compressed and used in many forms.

## Petroleum

Petroleum, or oil, is a naturally occurring liquid. Petroleum is *refined* or treated to make fuels and lubricants. Gasoline, fuel oil, jet fuel, and liquefied gases are made from petroleum. The Middle East has more than half of the world's known oil reserves. Other petroleum products made from crude oil include plastics, inks, tires, and pharmaceutical products (for medical uses).

**THINK GREEN:** Global Warming

# Ocean Acidification

Oceans cover more than 70% of Earth's surface. They are important in the carbon cycle. Oceans absorb a large amount of carbon dioxide from the atmosphere. More carbon dioxide in the atmosphere means more carbon dioxide in the oceans. Carbon dioxide dissolves in the ocean. It then reacts with water to form carbonic acid. This causes *ocean acidification*. The ocean water becomes more acidic.

# Ocean Acidification 53

## Ocean Water and the pH Scale

Oceans are a little less acidic than pure water. The pH scale measures the acidic or basic properties of any substance. The average value of seawater in the past was about 8.16. It was a little bit basic.

## What makes oceans acidic?

Carbon dioxide is released in large amounts by burning fossil fuels. About half of the carbon dioxide is absorbed by oceans. The oceans may have absorbed too much carbon dioxide. This has led to a decrease in the pH of Earth's oceans. This makes the ocean more acidic. Even a small change in pH may lead to large changes in ecosystem functioning.

## Effects on Marine Life

Higher ocean acidity means that there are more hydrogen ions in the water. These hydrogen ions combine with carbonate ions to form bicarbonate. This removes carbonate ions from the water. Then ocean organisms cannot use the carbonate to form their shells.

## Effects of Ocean Acidification

- In the Antarctic Ocean, the tiny-shelled planktons, which are the main food source of fish and other animals, suffer first.
- The high levels of carbon dioxide make it harder for fish and shellfish to breathe underwater.
- The chemical changes in the ocean water make the water less able to absorb carbon dioxide from the atmosphere. This speeds up global warming.

### Did you know?

By 2100, there may be a 150% increase in ocean acidification.

**THINK GREEN:** Global Warming

# Population Shifts of Plants and Animals

Plant and animal populations are affected by global warming. In the next 50 years, more than a million kinds of plants and animals will move toward extinction due to global warming. When a kind of plant or animal, or *species*, becomes extinct, there are no more living creatures of that kind.

## Moving Species

Scientists have discovered that about 2,000 species of plants and animals are moving toward the poles. During the second half of the 20th century, plant and animal species in mountain areas started moving up the mountains. They are looking for cooler temperatures.

Population Shifts of Plants and Animals    55

### Did you know?
If the global average temperature increases by more than 2.7°F to 4.5°F nearly 20% to 30% of all plant and animal species are likely to become extinct.

## Global Warming and Plant Shifts
- The spring wildflowers of southeastern New York are blooming earlier.
- In the Olympic Mountains in Washington, the trees of the forest near the base of the mountains have shifted. These trees now grow higher in the mountains.
- Less algae grows under the shrinking sea ice in the Antarctic Peninsula area.

## Global Warming and Animal Population Shifts
- In the past 30 years, dozens of frog species have come to the edge of extinction.
- In the Arctic, polar bears have to swim longer distances to reach ice floes. Some of them drown. The U.S. Geological Survey says that by the mid-21st century, two-thirds of the world's polar bear population groups will become extinct.
- Warmer ocean and air temperatures are resulting in the shift of shoreline sea life toward the north.

**THINK GREEN:** Global Warming

# Fire

Fire is a chemical reaction. The chemical reaction in fire produces heat and light, which we call flame. Three things are needed to create fire. These are oxygen, fuel, and heat. Hotter temperatures on Earth mean hotter and drier summers. Hot weather dries the forest floor. This creates the perfect conditions for fire. Large fires release tons of carbon dioxide into the atmosphere. This causes more global warming.

## Wildfire

Wildfires are unplanned and uncontrolled fires. Wildfires destroy farm fields, trees, houses, and other buildings. They may kill animals and people.

## Types of Wildfires

- Surface wildfires are the most common fires. They burn along the forest floor and move slowly. Surface wildfires burn and damage fallen trees, grasses, and shrubs.
- Ground wildfires occur easily. They usually start by lightning. They burn slowly below the forest floor. Ground fires burn tree roots, dead wood, and dry leaves.
- Crown fires are the most dangerous wildfires. They are spread by high winds and burn the tops of the trees.

## Health Hazards of Wildfire

- Burns
- Breathing problems
- Burning of lungs from breathing smoke or hot air
- Heart problems

## Drought and Wildfire

Warmer temperatures lead to drought conditions. Water evaporates more during the summers and fall. This makes the drought conditions worse.

### Did you know?

In 1997, the peatlands in Southeast Asia burned. They released about 2.67 billion tons of carbon dioxide.

## Ways to Prevent Fires

- Make fires away from trees or bushes. The blowing ash and cinder may cause a wildfire.
- Be ready to put out the fire quickly and completely.
- Never leave a fire burning, even a burning cigarette.

# CFLs

**C**ompact fluorescent lamps or CFLs are efficient alternatives to the older incandescent light bulbs. CFLs contain phosphorus, which gives soft light. These bulbs are 75% more energy-efficient and last 10 times longer than incandescent bulbs. CFLs also cut carbon dioxide emissions.

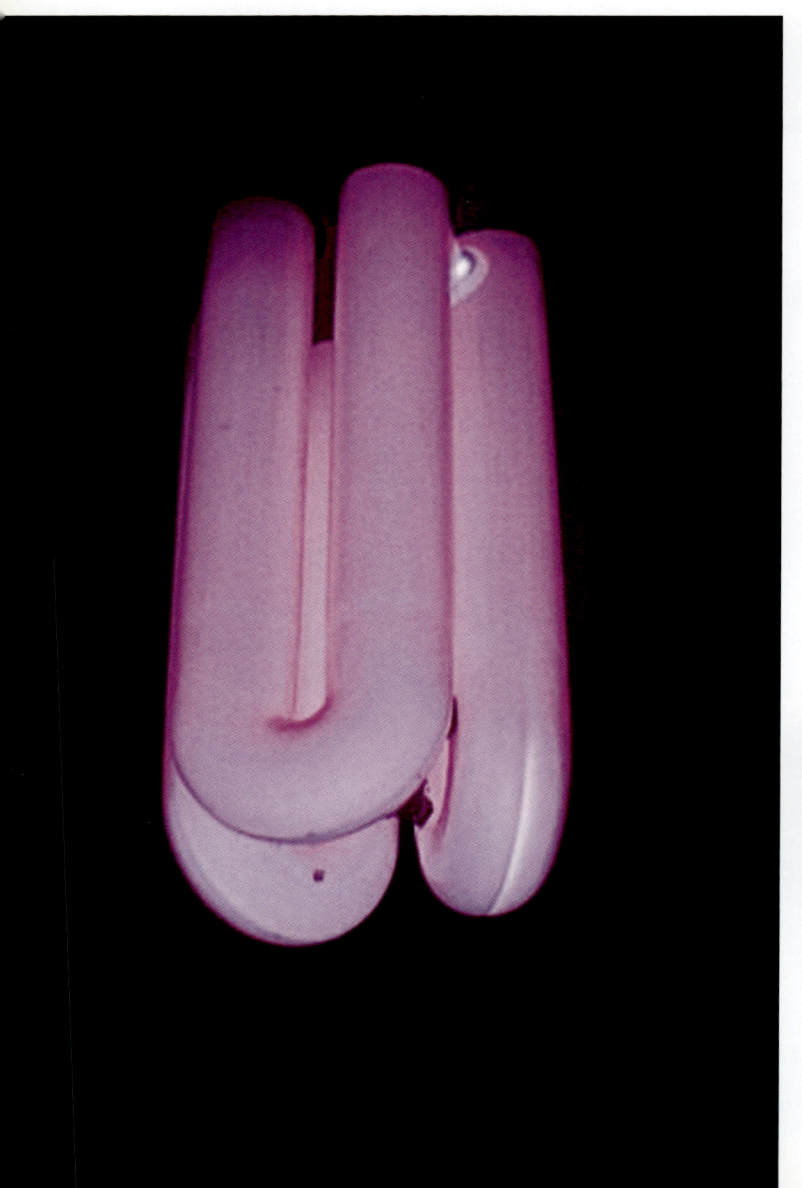

### CFLs and Global Warming
Global warming can be decreased by using energy-saving bulbs. CFLs save electricity. Less carbon dioxide, sulfur oxide, and high-level nuclear waste are created because less electricity is needed.

### CFLs Contain Harmful Mercury
Mercury is a chemical found naturally in the environment. CFLs contain a small amount of mercury. Mercury is harmful. Many environmental groups want makers to limit the amount of mercury in CFLs. They also want the proper disposal of CFLs. They don't want the mercury to become a health hazard.

## Advantages of CFLs

- Save energy
- Save about $25 to $45 over the life of the bulb
- Reduce emission of greenhouse gases
- Reduce water and air pollution
- Last up to 10 times longer than incandescent bulbs
- Produce 90% less heat

### Did you know?

The United States can save enough electricity to close 21 power plants if every person switches over five highly used light bulbs to compact fluorescents.

## Disadvantages of CFLs

- Give off a small amount of ultraviolet rays
- Contain mercury, which is harmful and needs proper disposal
- Don't work well in cold temperatures
- More expensive than incandescent bulbs
- Produce low light
- Larger than regular incandescent bulbs; may not fit in some fixtures

# Indoor Air Pollution

Indoor air pollution occurs within a house, building, or an enclosed area. Recent studies have shown that indoor air is more polluted than the outdoor air. Indoor air pollution is caused by mold, bacteria, chemicals, and other things.

| Pollutants | Sources | Health Effects |
|---|---|---|
| Radon | Ground under the building | Lung cancer |
| Lead | Lead-based paint, contaminated soil | Damages kidneys and the central nervous system |
| Formaldehyde | Furniture polished with formaldehyde; plywood | Headache, nausea, skin irritation, watery eyes |
| Toluene, phenols, ketones | Paints, varnishes, air fresheners, cleansers | Eyes-nose-throat (ENT) irritation, headache |
| Carbon dioxide, carbon monoxide, and oxides of sulfur | Combustion of fuels, tobacco smoking | Fatigue, dizziness, confusion |
| Suspended particles | Tobacco smoking, combustion of solid fuels | Nasal congestion, asthma, burning sensation in eyes |
| Pesticides | Moth and mosquito repellants and insecticides | ENT irritation, cancer, kidney failure |
| Biological pollutants (bacteria, mites, molds, pollens) | Wet or moist walls, ceiling, and carpets | Allergy, shortness of breath, ENT irritation |

## Sources of Indoor Air Pollution

Indoor air pollution comes from many different sources. Burned substances, such as coal, wood, oil, and tobacco, are the main sources of indoor air pollution. Building materials also add to indoor air pollution. Some contain asbestos, a material that doesn't burn but may harm health. *Bioaerosols* are very tiny organisms that can spread in the air. These include molds, bacteria, viruses, pollen, and dust mites. They pollute indoor air. *Radon* is a naturally occurring radioactive gas. It is a health hazard. Radon can enter homes through cracks in the foundation floor and other openings. Radon may increase the risk of cancer.

## Plants That Reduce Indoor Air Pollution
- Aloe vera
- Chinese evergreen, bamboo palm, and lily
- Chrysanthemum
- English ivy
- Spider plant

## How to Prevent Indoor Air Pollution
- Do not smoke indoors.
- Install fans in kitchens and bathrooms.
- Use safe, non-toxic cleaning products.
- Clean and dust your whole house regularly.
- Clean the air filters.
- Keep the indoor space free of moisture. Make sure to heat all rooms so moisture does not form on the walls of unheated areas.

## Indoor Air Pollution in Developing Countries
Developing countries are the most affected by indoor air pollution. Almost 90% of homes outside cities in developing countries depend on biomass fuel. Biomass fuels are wood, dung, and crop residues. People burn these fuels indoors in open fires or stoves.

## Health Effects of Indoor Air Pollution
- Asthma
- Cancer
- Dizziness
- Headache
- Irritation in eyes, nose, and throat
- Heart diseases
- Lung diseases

### Did you know?
About 15,000 to 21,000 lung cancer deaths in the United States are caused by radon.

**THINK GREEN:** Global Warming

# Facts and Figures

1. Scientists say that 69,000 square miles of Arctic ice, roughly equal to the size of Florida, has disappeared.
2. U.S. State Department studies suggest that each year, clearing and erosion cause a loss of forest areas four times the size of Switzerland.
3. Between 1990 and 2005, Brazil cleared an area of forest equal to the size of California.  
7. Montana's Glacier National Park now has only 27 glaciers. It had 150 in 1910.
8. By 2100, sea levels may rise between 7 and 23 inches. A rise of just 4 inches may leave many South Seas islands under water and flood large parts of Southeast Asia.
9. The sun has been getting hotter over the past 60 years.
10. The Adélie penguin population decreased by about 22% during the last 25 years.
4. Each year, about 20% of greenhouse gas emissions result from deforestation in developing countries.
5. About 4.4 million trees are cut down every day.
6. Each year about 1.6 billion trees are removed. Reforestation efforts plant only 0.6 billion trees. Almost 1 billion trees are not replaced.

11. About 99% of the energy used for transcontinental flights could be saved if people used video conferencing.
12. The United States could cut about 40% of its oil imports if only 1 in 10 Americans use public transportation daily.
13. By using public transportation, every American family could reduce household expenses by $6,200 per year.
14. The industrialized countries in the northern hemisphere give off about 90% of the chlorofluorocarbons in the atmosphere.
15. Each year, about 1.2 trillion gallons of untreated sewage, storm water, and industrial waste are added to the U.S. water supply.

**THINK GREEN:** Global Warming

# Index

**A**
Adélie penguin 62
Alaskan forest 33
algae 7, 15, 35, 55, 65
alkaline 53
alternative fuels 11, 13, 19, 38, 41
Antarctic Ocean 53
anthracite 51
Arctic 3, 5, 13, 19, 32–33, 42–43, 55, 62–63
argon 28
asbestos 60
atmosphere 4, 8–13, 15, 18, 24–26, 28–29, 37, 39, 42–43, 45, 48, 52–53, 56, 63, 65
Australia 17, 25, 47, 57

**B**
bacteria 60–61, 65
bioaerosols 60
biodiesel 38, 41
biomass 9
bituminous 51
Brazil 49
Britain 26

**C**
California 62
Canada 5
cardiovascular 57
chemicals 6, 12, 13, 15, 34, 35, 51, 60
Chicago 41
China 61
chlorine 11–12, 36
chlorofluorocarbons 10–13, 29, 63
chrysanthemum 61
Colorado 25
compressed natural gas 38, 41

convection currents 49
coral reefs 7

**D**
deforestation 3–4, 11, 15, 19, 48–49, 62–63, 65
Delhi Metro 41
droughts 4, 9, 57

**E**
ecosystems 7
emission 6, 11, 16, 29, 40, 51, 58–59, 63, 65
Energy Star 45
erosion 7, 33–35, 43, 62, 65
ethanol 38
Europe 6, 23
extinction 9, 54

**F**
fluorescent 11, 58
fluorine 11–12, 36
fossil fuels 34
France 6, 47
Fujita scale 20

**G**
gasoline 19, 36–41, 51, 63
greenhouse gases 4, 6, 8, 10–12, 16–17, 19, 24, 26, 29, 32, 42, 45, 51, 59
Greenland 33, 42

**H**
hail 20
halons 12
hibernation 54
Hungary 17

**I**
incandescent 11, 58, 59
incinerators 37
India 5, 17

Indian Ocean 26
Industrial Revolution 10, 11

**J**
Japan 17, 39

**K**
kibbutz 47
Kyoto 3, 16, 17

**L**
landslides 7
lignite 51
liquefied petroleum gas 38

**M**
methane 6, 8–11, 17, 29, 43, 51
methanol 38
mold 45, 60
Mueller, Derek 33

**N**
natural gas 4, 9–11, 38, 41, 45, 50–51
New York 41, 55
nitric acid 11
nitrous oxide 9
nonflammable 12
North Pole 32
nuclear power plants 35, 37

**O**
Olympic Mountains 55

**P**
phosphorus 58
photosynthesis 49
Poland 17
Polar animals 19

**R**
radioactivity 34
recyclable 11
Russia 17, 39, 42, 51

**S**
Saffir-Simpson scale 31
Siberia 42
smog 11, 37, 38
snowstorms 7, 20
solar energy 4, 19, 28
spider plant 61
storm surges 31
stratosphere 11, 12
sulfur dioxide 18, 37
sulfur hexafluoride 17
Switzerland 17, 62

**T**
thermostat 22
thunderstorms 20
tornadoes 20, 21, 22
troposphere 11

**U**
ultraviolet radiation 12, 33
United States 6–7, 10, 17, 20, 22, 25, 27, 31, 34–35, 38–39, 41, 45, 47–48, 51, 54, 57, 59–60, 62–63
U.S. Environmental Protection Agency 62

**V**
Vancouver 41
volatile organic compounds 37

**W**
wildfires 4, 7, 36, 57
wind energy 19
World Health Organization (WHO) 5

# Glossary

**acid rain**: rain containing sulfur dioxide and other pollutants in dissolved form
**agriculture**: growing and raising of crops and animals for food
**algae**: small, aquatic, rootless plants such as seaweed
**altitude**: the height of an area measured from sea level
**aquatic**: ability to live or grow in or on water
**atmosphere**: the layer of gases that surround the earth
**bacteria**: microscopic, single-celled organisms that can cause diseases
**carbon dioxide**: colorless gas in the Earth's atmosphere that helps in trapping heat close to the Earth
**climate**: the usual weather of an area
**commercial**: to do with buying or selling
**component**: different parts that combine with other parts to make up a whole
**condense**: to change from a vapor or gas to a liquid
**conservation**: protection or management of a valuable resource such as water
**contaminate**: to pollute by direct contact
**continent**: one of the seven large landmasses on the earth, such as Asia or North America
**crust**: the hard, outer layer of the earth
**decompose**: to break down and decay
**deforestation**: removal of forests by cutting of trees
**depletion**: the gradual loss of non-renewable sources
**dissolve**: to separate into molecules in a liquid
**drought**: a long period of little or no rain
**ecosystem**: a complex community of living things in a physical environment
**emission**: discharge of substances into the air
**endangered**: in danger of dying out or becoming extinct
**erosion**: wearing away of land or soil by water, wind, animals, and even human activity
**evaporation**: the process by which a liquid turns to a gas
**fertilizer**: a chemical substance used to improve soil and promote plant growth
**glacier**: a large body of ice that moves very slowly

**habitat**: an environment in which a plant or animal normally lives and grows
**heat energy**: a form of energy given off by the sun or another heat source
**hemisphere**: half of the Earth on either side of the equator
**hurricane**: a severe storm with winds of 74 mph or more
**hydrogen**: a light, colorless, and odorless gas that burns well
**infrastructure**: the basic facilities required by a community, such as roads, bridges, and pipelines
**insulation**: a material designed to prevent heat from escaping
**irrigation**: the addition of water to agricultural land using sprinklers, pumps, or pipes
**molecule**: the smallest particles of an element or compound
**nutrients**: substances that help living things be healthy and grow
**organic matter**: remains of dead plants and animals
**organism**: any living structure—plant, animal, fungus, bacterium—capable of growth and reproduction
**photosynthesis**: the process through which green plants use energy from sunlight to make their own food
**plankton**: very small plants and animals that float in water and drift with ocean currents
**pollutant**: a substance that contaminates and pollutes air, water, and land
**precipitation**: falling products of condensation of water vapor in the atmosphere, as in rain, snow, or hail
**radiation**: rays or waves of energy emitted from the sun
**reservoir**: a lake that is used to store water
**resource**: something ready for use
**sediment**: small particles such as sand or gravel that settles on the land or ocean floor
**toxic**: containing poison
**ultraviolet rays**: invisible light rays emitted from the sun
**urban**: related to city or town
**vapor**: the gaseous form of something that is usually liquid